Carlos Alberto Gomes de Souza

The use of the Delonix regia species

Carlos Alberto Gomes de Souza

The use of the Delonix regia species

Biological, medicinal and chemical applications and obtaining high added value products

ScienciaScripts

Imprint

Any brand names and product names mentioned in this book are subject to trademark, brand or patent protection and are trademarks or registered trademarks of their respective holders. The use of brand names, product names, common names, trade names, product descriptions etc. even without a particular marking in this work is in no way to be construed to mean that such names may be regarded as unrestricted in respect of trademark and brand protection legislation and could thus be used by anyone.

Cover image: www.ingimage.com

This book is a translation from the original published under ISBN 978-613-9-62510-9.

Publisher:
Sciencia Scripts
is a trademark of
Dodo Books Indian Ocean Ltd. and OmniScriptum S.R.L publishing group

120 High Road, East Finchley, London, N2 9ED, United Kingdom
Str. Armeneasca 28/1, office 1, Chisinau MD-2012, Republic of Moldova, Europe
Printed at: see last page
ISBN: 978-620-7-74582-1

SUMMARY

AUTHOR

2

Carlos Alberto Gomes de Souza

Chemist, Master in Natural Resources (Aproveitamentos de Recursos Naturais), Master Teacher, level M, Secretaria de Educaçâo do Cearà, Fortaleza, Cearà, Brazil

cagsouz@gmail.com

THANKS

To the government of the State of Ceará, and especially to the Ceará Education Department, for ensuring that I was able to take time off from teaching to continue my doctoral studies in Chemistry, and to all those who contributed directly and indirectly to the production of this work.

PRESENTATION

4

The species *Delonix regia* (Bojer ex Hook.) *Raf.*, popularly known as flamboyant, is an exotic forest species from Madagascar that is highly adapted to the environmental conditions of the tropical climate. The name flamboyant, which comes from France, means flaming, and is so called because of the very bright red color of its flowers. It has other popular names such as paradise flower, rosewood, queen of summer, sweetheart tree and red acacia. Due to its versatility, *Delonix regia* has an impressive range of biological, medicinal and chemical properties, and can also be used to produce high value-added products.

CHAPTER 1

Taxonomic and morphological characterization of the species *Delonix regia*

1.1 INTRODUCTION

Delonix regia (Figure 1.1) is an endangered species in the wild and endemic to the island of Madagascar. It is a medium-sized ornamental tree, widely planted in avenues and gardens in tropical and subtropical countries such as Brazil, India, Pakistan, Puerto Rico, etc. The plant has been voted one of the five most beautiful flowering trees in the world (SINGH & KUMAR, 2014).

Figure 1.*1-Delonix regia* **species**

Delonix regia is also known as the Royal Poinciana or Flamboyant. The species was previously placed in the genus *Poinciana*, in honor of Phillippe de Longvilliers de Poincy (1583-1660), who is considered to have introduced the species to the American continent. The name flamboyant, which is of French origin, means flaming, is so called because of the very bright red color of its flowers. It goes by other popular names such as flor-do-paraiso, pau-rosa, rainha do verão, àrvore dos namorados and acacia-rubra (ABDULLAHI & ABDULLAHI, 2005; MEENAKSHISUNDARAM, SANTHAGURU & RAJENDRAN, 2011).

1.2 TAXONOMIC CLASSIFICATION

Kingdom: Plantae

Division: Magnoliophyta

Class: Magnoliopsida

Order Fabales

Family: Leguminosae

Subfamily: Caesalpinioideae

Genre: *Delonix*

Species: *Delonix regia*

Botanical name: *Delonix regia* (Hook.) Raf.

1.3   ORIGIN AND GEOGRAPHICAL DISTRIBUTION

Delonix regia, popularly known as flamboyant, is a forest species endemic to the island of Madagascar. In its wild state, it is found in the west and north of Madagascar (Figure 1.2) and exotic in Brazil, Portugal, Cyprus, Egypt, India, Kenya, Niger, Sri Lanka, South Africa, Uganda, the United States of America, the Caribbean and so on, being highly adaptable to the environmental conditions of the world's tropical and subtropical climate regions (SINGH & KUMAR, 2014).

Figure 1.2-Geographical **distribution of** the *Delonix regia* species in Madagascar

1.4   AGROECOLOGY A

Its natural habitat is in the hot and humid tropical rainforest in the west and north of Madagascar. flamboyant requires a tropical or subtropical climate and adapts to various types of soil, from clay to sandy, adapting more easily to sandy soils. the best development of the plant occurs at annual temperatures between 14 and 26 °C, rainfall of more than 700 mm and altitude between 0 and 2000 m (SINGH & KUMAR, 2014). In regions with a warm climate, it grows rapidly, reaching 1.5 meters per year by the time it reaches adulthood. It grows in full sun or partial shade. It is a hardy tree and can tolerate drought, poor soil and salty conditions (LIM, 2012).

Depending on the region where it grows, D. regia can present itself as a deciduous or semi-deciduous tree, which means that in the coldest months of the year (fall/winter) or in times of drought, it loses its leaves; in some cases, only the branches and trunk or a few leaves remain. In this way, the plant prepares itself so that it doesn't lose water through evaporation via the leaves (PATRo, 2013).

1.5 MoRFoLoGY

The flamboyant tree (Figure 1.1) can reach a height of 15 to 20 m. Its crown is wide and umbrella-shaped. Its canopy is broad and umbrella-shaped, consisting of fine, delicate, lacy, fern-like foliage that can be taller than its own height.

Flowers

Figure 1.3- *Delonix regia* flowers

The flowers of the flamboyant (Figure 1.3) are large, orange-red or scarlet-red in color, with five petals; one of the petals, called the standard, is vertical and has traces of white and yellow and stands out because it is slightly larger than the others. The flowering season for this plant is March to July (ARoRA et al., 2010; JUNGALWALA & CAMA, 1962).

Leaves

Figure 1.4- *Delonix regia* leaves

Its leaves are feather-like, with a characteristic bright green color and are bipinnate. Each leaf is about 30-50 cm long and has 20 to 40 pairs of primary leaflets on it, each of which is divided into 10 to 20 pairs of secondary leaflets, as shown in Figure 1.4. (PACHECO-AGUIRRE et al., 2010; JAHAN et al., 2010; JAHAN et al., 2010).

al., 2010).

Ramos

The branches are horizontal, forming a diameter that is greater than the height of the tree, the canopy is umbrella-shaped and spreads into long branches (PACHECO-AGUIRRE et al., 2010; JAHAN et al., 2010).

Bark

The bark is smooth, gray-brown, with few fissures and many lenticels; the inner bark is light brown (PACHECO-AGUIRRE et al., 2010; JAHAN et al., 2010).

Madeira

The wood is soft and white in color (LAKSHMI,1987).

Roots

The roots are shallow, but quite aggressive, with part of them above the surface, making it unsuitable for decorating sidewalks, streets or near water pipes, sewers, walls and even electrical wiring. Its beauty is highlighted when planted singly or in small groups in large areas, such as parks, squares and extensive gardens in homes, factories and farms. The flamboyant has a moderate tolerance to salinity and can also be used on the coast (PACHECO-AGUIRRE et al., 2010; PATRO, 2013; NOGUEIRA et al., 2012).

Fruits

The husks (Figure 1.5-c) are smooth, grayish brown, slightly cracked and with many lenticular channels; the inner husks are light brown (PACHECO-AGUIRRE et al., 2010). The fruits, when young, are green and flaccid (Figure 1.5-a), turning dark brown (Figure 1.5-b) as they mature, becoming rigid woody pods, 30-50 cm long, 3.8 cm thick, 5-7.6 cm wide.

When ripe, the fruits (indehiscent) open naturally, showing many horizontally partitioned seed chambers, which finally open into two parts. The pods are harvested from August to October (ARORA et al., 2010).

(a) green fruit (b) ripe fruit (c) peel

Figure 1.5- *Delonix regia* fruits or pods

Seeds

The seeds are made up of husk, endosperm and embryo, as shown in Figure 1.6. They have a slight integumentary dormancy which can be broken by scarifying one end or immersing them in hot water (80°C) for 5 to 10 minutes. Germination takes place around two weeks after planting. Each pod has about thirty to forty-five seeds, which are hard, gray, shiny and about 2 cm long.

They have a rectangular shape and are very similar to date seeds, transversely spotted with a bony integument and weighing approximately 0.4 g. (ARORA et al., 2010; PATRO, 2013), as shown in Figure 1.6.

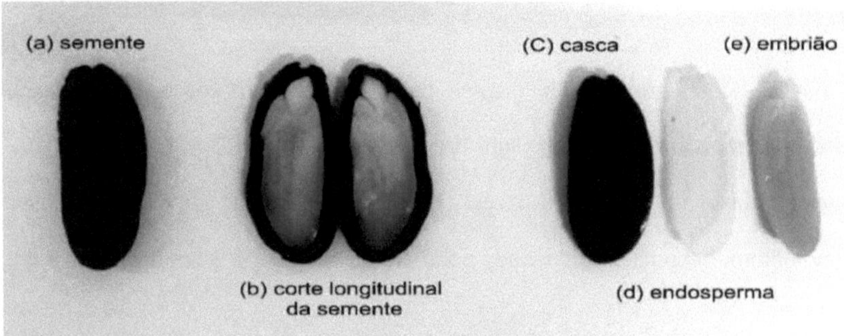

Figure 1.6- *Delonix regia* seed

REFERENCES

ARORA, A.; SEN, R. & SINGH, J. Fatty acid composition of *Delonix regia* (Gulmohar) seed oil from arid zone of Rajasthan. **Journal of Indian Council of Chemists**. v.27, p.150-152, 2010.

ABDULLAHI, S. A. & ABDULLAHI, G. M. Effect of boiling on the proximate, antinutrients and amino acid composition of raw *Delonix regia* seeds. **Nigerian food journal**. v. 23, p.128-132, 2005.

JUNGALWALA, F. B. & CAMA, H. R. Carotenoids in *Delonix regia* (Gul Mohr) flower. **Biochemical Journal**. v.85, p.1, 1962.

LAKSHMI, V. Constituents of wood of Delonix regia. **National Academy Science Letters**. v.10, p.197, 1987.

LIM, T. K. Garcinia gummi-gutta. In *Edible Medicinal And Non-Medicinal Plants* (p. 4555, 2012). Springer Netherlands.

MEENAKSHISUNDARAM, M.; SANTHAGURU, K. & RAJENDRAN, K. Effects of bioinoculants on quality seedlings production of *Delonix regia* in tropical nursery conditions. **Asian Journal of Biochemical and Pharmaceutical Research**. v. *1*, p.98107, 2011.

NOGuEIRA, N. W.; DE LIMA, J. S. S.; DE FREITAS, R. M. O.; RIBEIRO, M. C. C.; LEAL, C. C. P. & PINTO, J. R. D. S. Effect of salinity on the emergence and initial growth of flamboyant seedlings. **Journal of Seed Science**. v.34, p. 466-472, 2012.

PATRO, R. (Ed.). **Flamboyant- *Delonix regia*.** 2013. Available at:

<http://www.jardineiro.net/plantas/flamboyant-delonix-regia.html>. Accessed on: 29 Feb. 2016.

SINGH, S. & KUMAR, S. N. A review: introduction to genus *Delonix*. **World Journal of Pharmacy and Pharmaceutical Sciences**. v. *3*, p.2042-55, 2014.

CHAPTER 2

Biological, medicinal and chemical applications of the species *Delonix regia*

This chapter presents a collection of biological, medicinal and chemical activities that have been reported on *Delonix regia, which* are described below.

2.1 BIOLOGICAL APPLICATIONS

In recent years, the resistance of pathogenic microorganisms to a large number of drugs has increased due to the indiscriminate use of antimicrobials, which are commonly marketed and used to treat infectious diseases. This situation has forced researchers to search for new drugs. Plants are an excellent source for searching for new antipathogenic drugs, as flowers, leaves and bark usually contain the majority of a plant's active constituents (NOVAES et al., 2013).

Delonix regia contains active constituents such as polyphenolics, flavonoids, anthocyanins and phenolic acids as bioactive secondary metabolites, which are responsible for its antioxidant activity and correlated with medicinal uses, including anti-ulcer, anti-helminthic, hepatoprotective, antimicrobial, anti-rheumatic and anti-inflammatory activities (KHURSHEED et al., 2012; ADJÉ et al., 2010).

Phytochemical screening results show that the bark contains tannins, terpenoids, alkaloids, glycosides, carbohydrates and sterols (AHMED et al., 2011; JAHAN et al., 2010; JUNGALWALA & CAMA, 1962; SAMA, RAJA & YADAV, 2012; SHANMUKHA, PATEL & PATEL, 2011); the flowers contain tannins, saponins, flavonoids, steroids, alkaloids and carotenoids, while the leaves have alkaloids, glycoside, tannin and carbohydrates (AZAB, ABDEL-DAIM & ELDAHSHAN, 2013; AHMED et al.., 2011; BARUA & BARUA, 1966; CHAI et al., 2012; EL-SAYED et al., 2011; KUMAR, SHAIK & YESHWANTH, 2013; MUKHERJEE, 1975; NAJI & LUCA, 2013; PRAKASH et al., 2013; RAGASA & HOFILENA, 2011; RAHMAN et al..., 2011); the fruits and seeds contain fatty acids, amino acids, carbohydrates and tannins and sisterol (HOASAMANI & HOSAMANI, 1995; SUNG & FOWDEN, 1971; ADEWUYI et al., 2010; KAPOOR, 1972; TAMAKI, TERUYA & TAKO, 2010).

Table 2.1 shows the use of different parts of the plant with biological activity.

Table 2.1-Use of different parts of *Delonix regia* with biological activity.

Biological activity	Part of the plant	Reference
Antiarthritic	Flowers	CHITRA et al., 2010

Antidiabetic	Leaves	RAHMAN et al., 2011
Antidiarrheal	Flowers	SHIRAMANE et al., 2011
Antiemetics	Leaves	AHMED et al., 2012
Antifungal	Leaves	AHMAD & AQUIL, 2003; SHARMA et al., 2010 RANI, CHANDRAMOHAN & KUMARAVEL, 2012
Anthelmintic	Flowers	AHIRRAO et al., 2011
Anti-haemolytic	Flowers	VENNILA, 2013
Anti-inflammatory and analgesic	Flowers, leaves and bark	MURUGANANDAN et al, 2000; SHEWALE et al, 2012
Antimalarial	Flowers, leaves, pods, seeds, bark	FATMAWATY & ASTUTI, 2013
Antimicrobial	Flowers, leaves, seeds, stem, bark and root	RANI, CHANDRAMOHAN & KUMARAVEL, 2012; SAMA,RAGASA & HOFILENA, 2011; RAJA & YADAV, 2012
Antioxidant	Flowers, leaves and bark	RANI et al.,2011; SALEM, 2013; SHABIR et al., 2011; SHANMUKHA, PATEL & PATEL, 2011
Healing	Flowers	KHAN ET AL., 2012
Cytotoxic	Flowers, leaves and bark	RANJIT et al., 2014; PARUL, ALAM & RANA, 2014
Diuretic	Flowers	VELAN, 2016

		SHIRAMANE, CHIVDE & KAMSHETTY, 2011
Gastroprotective	Flowers	SHIRAMANE, CHIVDE & KAMSHETTY, 2011
Hepatoprotective	Flowers, leaves branches	andAHMED et al., 2011; EL-SAYED et al., 2011
Hypoglycemic	Leaves	SHEWALE et al., 2012; SHIRAMANE, CHIVDE & KAMSHETTY, 2011
Larvicide	Flowers	DEEPA & REMADEVI, 2011

2.2 MEDICINAL APPLICATIONS

Delonix regia has a wide range of medicinal properties and has been used in folk medicine in various civilizations. It is used to treat constipation, inflammation, arthritis, hemiplegia, leucorrhea, rheumatism, stomach ailments, bronchitis and pneumonia in children, fever, anemia, expectorant, abdominal pain, body and joint pain, flatulence, diarrhea, gastric problems, etc. (ABDEL-BARRY & AL- HAKIEM, 2000; CHATTERJEE & PAKRASHI, 1991; KHARE, 2008; SHABIR et al, 2011; SHIRAMANE et al., 2011).

In rural areas, its flowers are also used to prepare homemade teas which are traditionally known for their medicinal properties, such as antimicrobial and antifungal or antibiotic properties (CHANG-HUNG & LIH-LING, 1992; JOY et al., 2001; AQIL et al., 2005; JYOTHI et al., 2007).

2.3 CHEMICAL APPLICATIONS

In addition to its various biological and medicinal properties, there are reports that extracts of *Delonix regia* act as a corrosion inhibitor. Abiola and colleagues (2007) investigated the inhibitory effect of the species' extract in reducing the corrosion rate of aluminum in acidic media. The study aimed to find a low-cost and environmentally safe inhibitor to reduce the corrosion rate of aluminum.

Recently, it has been observed that the flowers of the plant are excellent adsorbents of heavy metals, increasing their versatility in their applications Jimoh and collaborators (2012), for example, showed the ability of the flowers of the plant to remove the ions of Co (II), Cu (II) and Pb (II), by adsorption in aqueous solutions analyzed in multiple metal experiments at 32 °C. This study showed that flamboyant flowers are suitable agricultural waste for the removal of Co (II), Pb (II) and Cu (II) from an ionic aqueous solution and that they could be used as an alternative to activated charcoal in the removal of metal ions in aqueous effluents (JIMOH et al., 2012).

In other studies the bark has been used as a biomonitor and bioaccumulator of atmospheric trace metals (UKPEBOR et al., 2010), as well as for the removal of Ni from aqueous solutions through adsorption (PATIL & SHRIVASTAVA, 2010).

The 5% chloroform extract of the bark tissue showed potential as a termiticide agent with 80% mortality in 48 h against termites (RUPAL, SAVALIA & NARASIMHACHARYA, 2011).

REFERENCES

ABDEL-BARRY, J. A. & AL-HAKIEM, M. H. Acute intraperitoneal and oral toxicity of the leaf glycosidic extract of Trigonella foenum-graecum in mice. **Journal of ethnopharmacology**. v. *70*, p.65-68, 2000.

ADEWUYI , A., ODERINDE, R. A., RAO, B. V. S. K., PRASAD, R. B. N. & ANJANEYULU, B. Chemical component and fatty acid distribution of Delonix regia and Peltophorum pterocarpum seed oils. **Food science and technology research.** v. *16*, p.565-570, 2010.

ADJÉ, F., LOZANO, Y. F., LOZANO, P., ADIMA, A., CHEMAT, F. & GAYDOU, E. M.

(2010 Optimization of anthocyanin, flavonol and phenolic acid extractions from Delonix regia tree flowers using ultrasound-assisted water extraction. **Industrial Crops and Products.** v. *32*, p.439-444, 2010.

AHIRRAO, R. A., PATEL, M. R., HAMID, S. & PATIL, J. K. In vitro anthelmintic property of Gulmohar flowers against Pheritima posthuma. **Pharmacologyonline.** v. *1*, p.728732, 2011.

AHMAD, I. & AQUIL, F. Broad spectrum antibacterial and antifungal activities and potency of crude, alcoholic extract and fraction of *Delonix regia* flowers. In *2nd World Congress on[a] Biotechnological Development of Herbal Medicine'''*, NBRI, Lucknow, UP, India (p. 74), 2003.

AHMED, S., SABZWARI, T., HASAN, M. M. & AZHAR, I. ANTIEMETIC ACTIVITY OF LEAVES EXTRACTS OF FIVE LEGUMINOUS PLANTS. **International Journal of Research in Ayurveda & Pharmacy.** v. *3*, 2012.

AHMED, J., NIRMAL, S., DHASADE, V., PATIL, A., KADAM, S., PAL, S. & PATTAN, S. Hepatoprotective activity of methanol extract of aerial parts of Delonix regia. **Phytopharmacology.** v. *1*, p.118-122, 2011.

AQIL, F.; KHAN, S.M.A.; OWAIS, M. & AHMAD, I. Effect of certain bioactive plant extracts on clinical isolates of beta-lactamase producing methicillin resistant *Staphylococcus aureus.* **Journal of Basic Microbiology**. v. 45, p. 106-114, 2005.

AZAB, S. S., ABDEL-DAIM, M. & ELDAHSHAN, O. A. Phytochemical, cytotoxic, hepatoprotective and antioxidant properties of *Delonix regia* leaves extract. **Medicinal Chemistry Research**. v. 22, 4269-4277, 2013.

BARBOSA-FILHO, J. M., PIUVEZAM, M. R., MOURA, M. D., SILVA, M. S., LIMA, K. V. B.,

DA-CUNHA, E. V. L. & TAKEMURA, O. S. Anti-inflammatory activity of alkaloids: A twenty-century review. **Revista Brasileira de Farmacognosia.** *v.16*, p.109-139, 2006.

BARUA, R. K. & BARUA, A. B. Oxidation of zeaxanthin: Isolation and properties of 3-hydroxyretinene. **Biochemical Journal**. v. *101*, p.250, 1966.

CHAI, W. M., SHI, Y., FENG, H. L., QIU, L., ZHOU, H. C., DENG, Z. W., YAN, C.L. & CHEN, Q. X. NMR, HPLC-ESI-MS, and MALDI-TOF MS analysis of condensed tannins from Delonix regia (Bojer ex Hook.) Raf. and their bioactivities. **Journal of agricultural and food chemistry**. v. *60*, p.5013-5022, 2012.

CHANG-HUNG, C. & LIH-LING, L. Allelopathic substances and interactions of Delonix regia(Boj) Raf. **Journal of Chemical Ecology**. v.18, p. 2285-2303, 1992.

CHATTERJEE, A. & PAKRASHI, S. C. The treatise on Indian medicinal plants: vol. 1. *New Delhi: Publications and Information Directorate, CSIR 172p.-illus., col. illus. ISBN 8172360118 En Icones. Includes authentic Sanskrit slokas in both Devnagri and Roman scripts. Plant records. Geog, 6, 1991.*

CHANG-HUNG, C. & LIH-LING, L. Allelopathic substances and interactions of *Delonix regia*(Boj) Raf. **Journal of Chemical Ecology**. v.18, p. 2285-2303, 1992.

CHITRA, V., ILANGO, K., RAJANANDH, M. G. & SONI, D. Evaluation of Delonix regia Linn. flowers for antiarthritic and antioxidant activity in female wistar rats. **Annals of Biological Research**. v. *1*, p.142-147, 2010.

DEEPA, B. & REMADEVI, O. K. Larvicidal activity of the flowers of *Delonix regia* (Bojer Ex Hook.) Rafin. (Fabales: Fabaceae) against the Teak defoliator, *Hyblaea puera* Cramer. **Current Biotica**. v. *5*, p.237-240, 2011.

EL-SAYED, A. M., EZZAT, S. M., SALAMA, M. M. & SLEEM, A. A. Hepatoprotective and cytotoxic activities of *Delonix regia* flower extracts. **Pharmacognosy journal.** v. *3*, p.49-56, 2011.

FATMAWATY, F. & ASTUTI, H. Antimalarial activity of *Delonix regia* on mice with *Plasmodium berghei*. **Journal of Natural Products**. v. *6*, p.61-6, 2013.

HOASAMANI, K. M. & HOSAMANI, S. K. Component Fatty Acids of Delonix regis Seed Oil-ASource of7-(2-Octacyclopropen-1-yI) heptanoic Acid and 8-(2-OctacycIopropen-1-yl) octanoic Acid. **European Journal of Lipid Science and Technology**. v. *97*, p.420-422, 1995.

JAHAN, I., RAHMAN, M., RAHMAN, M., KAISAR, M., ISLAM, M., WAHAB, A. & RASHID, M. Chemical and biological investigations of *Delonix regia* (Bojer ex Hook.) Raf. **Acta Pharmaceutica**. v. *60*, p.207-215, 2010.

JOY, P.P.; THOMAS, J.; MATHEW, S. & SKARIA, B.P. Medicinal Plants. In: Tropical Horticulture Vol.2 Ed: T.K. Bose, Naya Prakash, Culcutta, J. Kabir and P.P. Joy, p. 449632, 2001.

JUNGALWALA, F. B. & CAMA, H. R. Carotenoids in *Delonix regia* (Gul Mohr) flower. **Biochemical Journal** v. *85*, p.1, 1962.

JYOTHI, M.V.; MANDAYAN, S.N.; KOTAMBALLI, N.C.; & BHAGYALAKSHMI, N. Antioxidative efficacies of floral petal extracts of *Delonix regia* Raf. **International Journal of Biomedical and Pharmaceutical Science**, v.1, p.73-82, 2007.

KAPOOR, V. P. A galactomannan from the seeds of *Delonixregia*. **Phytochemistry**, v. 11, p. 1129-1132, 1972.

KHAN, M. A., SAXENA, A., FATIMA, F. T., SHARMA, G., GOUD, V. & HUSAIN, A. Study of wound healing activity of *Delonix regia* flowers in experimental animal models. **Am J PharmTech Res**, v. *2*, p.380-90, 2012.

KHARE, C. P. *Indian medicinal plants: an illustrated dictionary*. Springer Science & Business Media, 2008.

KHURSHEED, R., NAZ, A., NAZ, E., SHARIF, H. & RIZWANI, G. H. Antibacterial, antimycelial and phytochemical analysis of ricinus communis linn, trigonella foenum grecum linn and *Delonix regia* (Bojer ex Hook.) Raf of Pakistan. **Romanian Biotechnological Letters**, v. *17*, p.7237-7244, 2012.

KUMAR, A. R., SHAIK, R. & YESHWANTH, D. Phytochemical evaluation of *Delonix regia*, *Samanea saman* and *Bauhinia variegata*. **International journal of research in pharmacy and chemistry**. v. *3*, p.768-772, 2013.

MUKHERJEE, D. Keto and imino acids in *Delonix regia* flowers. **Phytochemistry**. v. *14*, p.1915-1918, 1975.

MURUGANANDAN, S., SRINIVASAN, K., TANDAN, S. K., LAL, J., CHANDRA, S. & RAVIPRAKASH, V. Anti-inflammatory and analgesic activities of some medicinal plants. **Anti-inflammatory and analgesic activities of some medicinal plants**. v. *22*, p.56-58, 2000.

NAJI, E. & LUCA, R. Comparative study of anthocyanins content of some selected Yemeni plants. **Journal of Pharmacy and Phytotherapeutics**. v. *1*, p.5, 2013.

NOVAIS, T. S., COSTA, J. F. O., DAVID, J. P. D. L., DAVID, J. M., QUEIROZ, L. P. D., FRANÇA, F., GIULIETTI, A.M.IV, V; SOARES, M.B.P. & SANTOS, R. R. D. Antibacterial activity in some plant extracts from the Brazilian semi-arid region. **Revista Brasileira de Farmacognosia**. v. *13*, p.5-8, 2003.

PARUL, R., ALAM, J. & RANA, S. Antinociceptive and cytotoxic potential of ethanolic extract of *Delonix regia* (Leaves), 2014.

PATIL, A. K. & SHRIVASTAVA, V. S. Adsorption of Ni (II) from aqueous solution on *Delonix regia* (Gulmohar) tree bark. **Archives of Applied Science Research**. v.2, p.404-413, 2010.

PRAKASH, M., GOVINDSWAMY, C., RAJU, R. & BEEVI, F. Isolation and antibacterial activity of oleananoic acid acetate from *Delonix regia* leaves. **Journal of Pharmacy Research**, v. 6, p.423-425, 2013.

RAGASA, C. Y. & HOFILENA, J. G. Antimicrobial coumarin derivative from *Delonix regia*. **The Manila Journal Science,** p. 7, p. 7-11,2011.

RAGHUNATHAN, K. & MITRA, R. **Pharmacognosy of indigenous drugs** (Vol. 1).

Central council for research in Ayurveda and Siddha, (Eds.)1999.

RAHMAN, M., HASAN, N., DAS, A. K., HOSSAIN, T., JAHAN, R., KHATUN, A. & RAHMATULLAH, μ. Effect Of *Delonix Regia* Leaf Extract On Glucose Tolerance In Glucoseinduced Hyperglycemic Mice. **African Journal of Traditional, Complementary and Alternative Medicines**. v. 8, 2011.

RANI, P. M.J., KANNAN, P. S. M. & KUMARAVEL, S. Screening of antioxidant activity, total phenolics and gas chromatograph and mass spectrometer (GC-MS) study of *Delonix regia*. **African journal of biochemistry research**. v.2, p.341-347, 2011.

RANI, J. M. J., CHANDRAMOHAN, G. & KUMARAVEL, S. Evaluation of antimicrobial activity of some garden plant leaves against *Lactobacillus Sp, Streptococcus mitis, Candida albicans* and *Aspergillus niger*. **African Journal of Basic & Applied Sciences**. v. 4, p.139-142, 2012.

RANJIT, P. M., TEJASWI, J., KUMAR, K. P., ANKAMMA, Y. C. & GIRIJASANKAR, G. In vitro Cytotoxic Activity of Hydro Ethanolic Extract of *Delonix regia* (Bojer ex. Hook.) Flowers on Cancer Cell Lines. **British Journal of Pharmaceutical Research**. v.*4*, p.443-452, 2014.

RUPAL, A. V., SAVALIA, D. M. & NARASIMHACHARYA, A. V. R. L. PLANT EXTRACTS AS BIOTERMITICIDES. **Electronic Journal of Environmental Sciences**. v. *4*, 2011.

SALEM, M. Z. Evaluation of the antibacterial and antioxidant activities of stem bark extracts of *Delonix regia* and *Erythrina humeana* grown in Egypt. **Journal of forest products and industries**. v.*2*, p. 48-52, 2013.

SAMA, K., RAJA, A. & YADAV, R. H. Antibacterial and pharmacognostical evaluation of *Delonix regia* root bark. **International Journal of Pharmacy & Life Sciences**. v.*3*, 2012.

SEETHARAM, Y. N., SHARANABASAPPA, V. G., MURTHY, N. S. & SANGAMMA, Y.

R. Antimicrobial and analgesic activity of *Delonix elatagamble* and *Delonix regia*. *Raf.* **Aryavaidyan** .v.*16*, p.56-3, 2002.

SHABIR, G., ANWAR, F., SULTANA, B., KHALID, Z. M., AFZAL, M., KHAN, Q. M. & ASHRAFUZZAMAN, M. Antioxidant and antimicrobial attributes and phenolics of different solvent extracts from leaves, flowers and bark of Gold Mohar [*Delonix regia* (Bojer ex Hook.) Raf.]. **Molecules**. v.16, p.7302-7319, 2011.

SHANMUKHA, I., PATEL, H. & PATEL, J. Riyazunnisa. Quantification of total phenol and flavonoid content of *Delonix regia* flowers. **International Journal of ChemTech Research**. v. *3*, p.280-283, 2011.

SHARMA, R. A., CHANDRAWAT, P., SHARMA, S., SHARMA, D., SHARMA, B. & SINGH, D. Efficacy of *Delonix regia* rafin (syn. *Poinciana regia* bojer ex. Hook) for potential antifungal activity.

Bioscan, v. *5*, p.441-444, 2010.

SHEWALE, V. D., DESHMUKH, T. A., PATIL, L. S. & PATIL, V. R. (2012). Antilnflammatory Activity of Delonix regia (Boj. Ex. Hook). *Advances in pharmacological sciences, 2012*.

SHIRAMANE, R., CHIVDE, B. & KAMSHETTY, м. Gastroprotective activity of ethanolic extract of Delonix regia flowers in experimental induced ulcer in wistar albino rats. **International journal of research in pharmacy and chemistry**. v.2, p.234-8, 2011.

SHIRAMANE, R. S., BIRADAR, K. V., CHIVDE, B. V., SHAMBHULINGAYYA, H. M. & GOUD, V. In-vivo antidiarrhoeal activity of ethanolic extract of Delonix regia flowers in experimental induced diarrhoea in wistar albino rats. **International Journal of Research in Pharmacy and Chemistry**. v. *1*, p.2231-2781,2011.

SUNG, M. L. & FOWDEN, L. Imino acid biosynthesis in Delonix regia. **Phytochemistry**. v. *10*, p.1523-1528, 1971.

TAMAKI, Y., TERUYA, T. & TAKO, M. The chemical structure of galactomannan isolated from seeds of *Delonix regia*. **Bioscience, biotechnology, and biochemistry**. v. *74*, p.1110-1112, 2010.

UKPEBOR, E. E., UKPEBOR, J. E., AIGBOKHAN, E., GOJI, I., ONOJEGHUO, A. O. & OKONKWO, A. C. *Delonix regia* and *Casuarina equisetifolia* as passive biomonitors and as bioaccumulators of atmospheric trace metals. **Journal of Environmental Sciences**. v. *22*, p.1073-1079, 2010.

VENNILA J. A study on antihemolytic activity of flavonoid fraction from the petals of *Delonix regia*. **International journal of ethnomedicine and pharmacological research**. v.1, p.15-20, 2013.

VELAN, S. S. Evaluation of Diuretic activity of *Delonix regia* (Gul Mohr) flowers in Albino rats. **International Journal of Research in Pharmaceutical Sciences**, v. 3, p.369-372, 2016.

CHAPTER 3

Using the *Delonix regia* species to obtain high added value products

Delonix regia contains proteins, flavonoids, tannins, phenolic compounds, glycosides, sterals and terpenoids. In addition, there are reports on the isolation of noble substances extracted from its seeds, galactomannans and fatty acids; from its flowers and leaves, substances with pharmacological potential; and on the use of its pods in the production of activated charcoal. In order to obtain high value-added products from the plant, certain factors must be taken into account, such as: flowering and fruiting season, climate and soil types, since they influence the availability of raw materials.

3.1 obtaining chemical products

There are few reports of the use of *Delonix regia* biomass residues in the production of chemical products. In one of these studies, Grecco and colleagues (2016) studied the production of fuels and chemical inputs, especially aromatic compounds, using zeolitic catalysts (USY and Ga/USY), through the fast pyrolysis of biomass from the pod husk of the plant. It was observed that, in the absence of the catalyst, the fast pyrolysis of the different biomasses produced mainly oxygenated and nitrogenous compounds. On the other hand, in the presence of the USY zeolite, mainly aromatic compounds were obtained. In addition, the incorporation of gallium decreased the production of oxygenated and nitrogenous compounds and increased the yield of aromatic compounds, especially o-xylene, a high value-added chemical product used as a starting material in various industrial processes (GRECCO et al., 2016).

3.2   OBTAINING ACTIVATED CARBON

Conventionally, the main precursor materials for the preparation of activated carbons are oil residues, wood, peat (fossil coal) and stone coal (embers) (OLIVEIRA et al., 2008). However, in recent years there has been growing interest in the production of activated carbon from biomass, due to its low cost and the reduction of waste (JUAN & QIANG, 2009).

A wide variety of lignocellulosic biomasses have been widely used as precursor materials in the production of activated carbon, such as guava seeds (GONZALEZ & MONTOYA, 2009), almond shells (NABAIS et al., 2011), coffee rejects (PEREIRA, et. al., 2008), sorghum kernels (DIAO et al., 2002), walnut shells (ZABIHI et al., 2010; HAYASHI et al., 2002), rice husks (KUMAGAI et al., 2009), olive stones (YAVUZ et al., 2010), palm shells (DAUD & ALI, 2004), peanut shells (GIRGIS et al., 2002; DA SILVA, 2013), pistachio shells (YANG & LUA, 2003), sugarcane bagasse (KALDERIS et al., 2008), coconut shells (LIMA et al., 2013; LIMA , 2013).

The pods of *Delonix regia*, which gradually fall from the trees after a period of ripening, have already been used to obtain activated carbon, indicating that they can be a very important precursor in the production of high value-added products (IDRIS et al, 2011;

VARGAS et al., 2012; SUGUMARAN et al., 2012; VARGAS et al., 2011; VARGAS et al., 2011; VARGAS et al., 2010).

Considering that the composition of biomass can vary widely, it is necessary to know its centesimal, elemental and biochemical composition before using it to produce charcoal. In this way, SUGUMARAN and colleagues (2012) characterized the plant's pod for the production of activated carbon, using chemical activation with phosphoric acid and potassium hydroxide. The results of the centesimal, elemental and biochemical analysis of the sample are shown in Table 3.1.

Table 3.1- Immediate, elemental and biochemical analysis of *Delonix regia* pods.

Immediate analysis (%)			
Moisture content	Ash content	Volatile materials	Fixed carbon
0,22±0,04	2,80±0,65	92,03±3,74	5,20±3,81

Elemental analysis (%)				
Carbon	Hydrogen	Nitrogen	Sulphur	Oxygen
34, 224,	50	1,94	0,42	58,91

Biochemical analysis (%)		
Cellulose	Hemicellulose	Lignin
13,90	24,13	23,36

Source: adapted by SUGUMARAN et al., 2012

As expected, the lignocellulosic residue has a high carbon and oxygen content, which are the basic constituents of lignin, cellulose and hemicellulose. The ash content of the pod (2.80 %) is low, indicating that this raw material is suitable for making activated charcoal. This information is of great value, since ash plays an important role in adsorption processes, modifying the interaction between the carbon surface and the adsorbate molecule (SUGUMARAN et al., 2012).

Furthermore, in processes involving supported catalysts, ash can modify the behavior of the phases through electronic or structural interactions (RODRÎGUEZ-REINOSO and SEPÛLVEDA-ESCRIBANO, 2001).

The activated carbons produced by the pods of the flamboyant fruit have served as the basis for studies into efficient adsorbents for removing dyes such as methylene blue (OWOYOKUN, 2009; PONNUSAMI et al., 2009; VARGAS et al., 2011; SUGUMARAN et al., 2012). Methylene blue is used as a model compound in the removal of organic and colored contaminants from aqueous solutions (HAMEED et al., 2007).

Other dyes have also been studied using charcoal obtained from plants. Vargas et al. (2012), for example, used activated carbon obtained from *Delonix regia* to remove the dyes acid yellow 6, acid yellow 13 and acid red 18, either alone or in combination. They are responsible for the coloring of many products for human consumption (AL-DEGS et al., 2011).

In addition to studies related to dye adsorption, activated charcoal from the pod of the species has been used to adsorb heavy metals such as mercury. Ramakrishnan and Nagarajan (2008), for example, used charcoal from the pod to remove mercury, Hg (II), from aqueous effluents. The results showed that the adsorption process follows a pseudo-second order kinetic model. Around 92% of the adsorbed mercury (II) was recovered by washing the charcoal with distilled water and a 3% solution of potassium iodide.

3.3 BIOPOLYMER OBTAINING

Seeds have been considered an alternative to exploiting forests because they generate income for communities without causing a major impact on nature. It is important to know the level of exploitation that each species can withstand without damaging its reproduction (flowers and fruit) and regeneration (new plants being born) (ZANI et al., 2013). When these limits are respected, we have what is known as "sustainable management" of the species (NUNES & VIVIAN, 2011).

Respecting "sustainable management", basic studies on the chemical composition of tree species seeds are becoming progressively necessary as these products need to be used as a food resource for animals and humans (ZANI et al., 2013). In addition, these seeds can serve as an alternative source for the agrochemical industries that produce oils, and also for those that are currently producing green fuel - biodiesel (VALLILO et al., 2007).

Galactomannan is the name given to neutral polysaccharides extracted from the endosperm of seeds of certain legumes. Since ancient times, these polymers, often called gums, have been extracted or isolated from the seeds of carob trees, locust trees (lamb's lettuce), guar plants, taras (false brazilwood) and tamarind trees (LAPASIN and PRICL, 1999). Galactomannans stand out as extremely important ingredients in the food industry, since they result in solutions with high viscosity, act as emulsifiers and interact effectively with other polysaccharides to form gels.

Gums extracted from seeds have the ability to modify the behaviour of water in food systems in a highly efficient way, to reduce and minimize friction in foods (aiding in the processing and palatability of products) and to help control the size of crystals in saturated sugar solutions (STEPHEN and CHURMS, 1995).

Gums extracted from seeds have the ability to modify the behaviour of water in food systems in a highly efficient way, to reduce and minimize friction in foods (aiding in the processing and palatability of products) and to help control the size of crystals in saturated sugar solutions (STEPHEN and CHURMS, 1995).

These polysaccharides are polymers made up of a skeleton made up of repeated β-D-mannose units linked together by 1→4-type oxygen bridges. The α-D-galactose units are linked to the main chain by oxygen bonds of the 1→6 type (Figure 3.1). The content and distribution of D-galactose units depend on the origin and species of legume, and the extraction techniques used to obtain them (AZERO and ANDRADE, 1999).

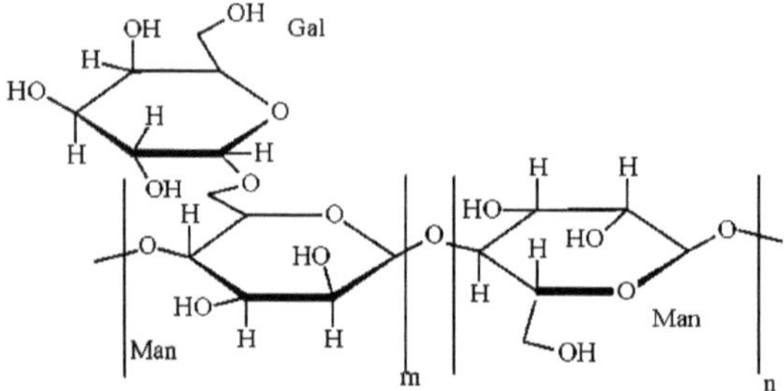

Figure 3.1-Partial **structure** of galactomannan

With regard to the distribution sequence of the D-galactose units along the mannose chain, the polymer can be classified as randomly distributed, alternating or in blocks, a factor which determines the interaction with other polysaccharides (DEA & MORRISON, 1975). Figure 3.2 shows, in schematic form, the distribution sequence of the D-galactose units along the D-mannose chain.

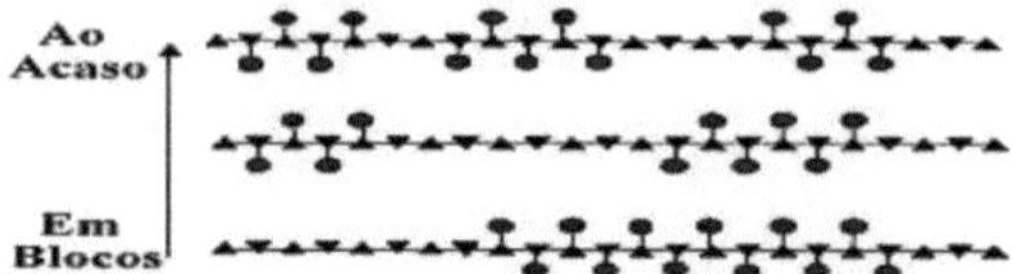

Figure 3.2- Sequence of distribution of D-galactose units in galactomannans Source: adapted from POLLARD & FISCHER, 2006.

The production of galactomannan is a lucrative activity in many regions of the world and involves several stages, ranging from planting and harvesting the seeds to obtaining the gum industrially.

The first step in evaluating a new product is to study its functional properties, which is usually done through its rheological characterization.

Exalbuminous seeds are made up of husk, endosperm and embryo, as shown in Figure 3.3:

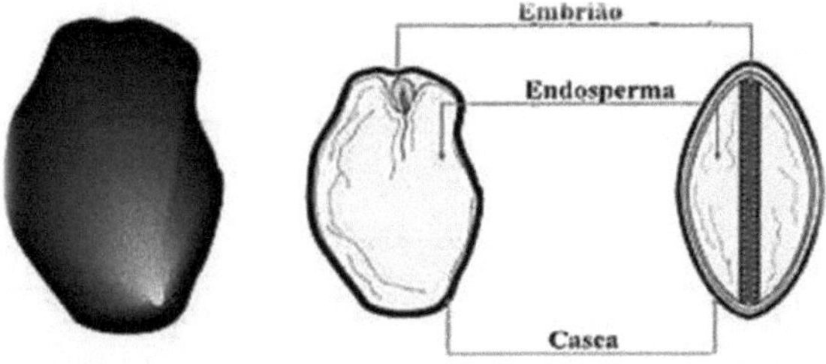

Figure 3.3- Longitudinal section of carob seed (*Ceratonia siliqua L.*) from the Leguminosae family containing endosperm

Source: adapted from DAKIA *et al.*, 2008.

In the germination of various seeds belonging to the Leguminosae family, galactomannan, which is the reserve polysaccharide of the endosperm, is degraded by 3 types of enzyme: α-galactosidase, β-mannanase and β-mannoside hydrolase to release energy (MCCLEARY, 1984).

Most galactomannans extracted from higher plants originate from the Leguminosae or Fabaceae family (DEA and MORRISON, 1975). According to Prajapati et al. (2013) around 70 species of Leguminosae have been identified as containing galactomannan. Traditionally, the Leguminosae family is subdivided into 3 subfamilies: Papilionoideae or

Faboideae, Caesalpinioideae and Mimosoideae (LOPES *et al.*, 2009).

Galactomannan is a versatile material and is used in many applications. They are excellent emulsion stabilizers and because they are non-toxic, they can be used in the textile, pharmaceutical, biomedical, cosmetics and food industries (BAVEJA *et al.*, 1991; KRISHNAIAH et *al.*, 2002; VARSHOSAZ *et al.,* 2006; VENDRUSCOLO et al., 2009; VIEIRA et *al.,* 2007).

As they can be extracted from various vegetable sources, galactomannans can differ in their molecular weight and molecular weight distribution. The main difference between galactomannans from different seed species is the mannose: galactose ratio (AQUILA et al., 2012). Table 3.2 shows the differences between galactomannans extracted from the seeds of some species found in the Brazilian flora.

Table 1.2- Ratio of mannose (M) to galactose (G) in various species of the Leguminosae family from the Brazilian flora.

Classification Botanic		Popular name	M/G ratio	References
Subfamily	Species			
Caesalpiniacae	*Senna multijuga*	Hallelujah	2,3: 1,0	RECHIA *et al.*, 1995
	Schizolobium amazonicum	Paricà	3,0: 1,0	GANTER *et al.*, 1995
	Schizolobium parahybae	Guapuruvu	3,0: 1,0	GANTER *et al.*, 1995.
	Cassia fastuosa	Mari-mari da mata	4,0: 1,0	MERCÊ *et al.*, 1998
	Caesalpinia pulcherrima	Cockroach beard	3,1:1,0	AZERO and ANDRADE, 1999
	Dimorphandra mollis	Beech	2,7: 1,0	PANEGASSI et al., 2000

	Cassia javanica	Matacaladorana	3,23: 1,0	AZERO and ANDRADE, 2002
	Dimorphandra gardneriana	Fava d'anta	1,84: 1,0	CUNHA *et al.*, 2009
	Caesalpinia ferrea	Jucà	2,1: 1,0	DE SOUZA *et al.*, 2010
	Delonix regia	Flamboyant	4,28: 1,0	BENTO *et al.*, 2013; DE SOUZA et al., 2015
Mimosoideae	*Mimosa scabrella*	Bracatinga	1,1: 1,0	GANTER *et al.*, 1995
	Stryphnodendron barbatiman	Barbatimâo	1,5: 1,0	GANTER *et al.*, 1995

Source: De Souza, 2018.

In the food industry, galactomannans are used as components of dairy products, fruit-based water gels, powdered products, breads, dietary products, coffee, baby milk, seasonings, sauces, soups and canned meats. The wide variety of applications reflects a large number of different functional characteristics, including high viscosity in solution form, stability in frozen systems and gel formation when mixed with other polysaccharides and proteins (GIDLEY and REID, 2006).

The seed gum of the *Delonix regia* species is a polysaccharide that belongs to the galactomannan group. Normally, these natural polymers have, as their main chain, repetitive β-D-mannopyranose units linked together by hydrogen bridges of the 1-4 type and α-D-galactopyranose side chains linked together of the 1→6 type (mannose/galactose ratio 2:1). The mannose/galactose ratio is similar to that of guar gum, but they differ in terms of the position of the OH bond in the main chain: flamboyant gum has α-D-mannose, while guar gum has β-D-mannose (Figure 3.1) (KAPOOR, 1972).

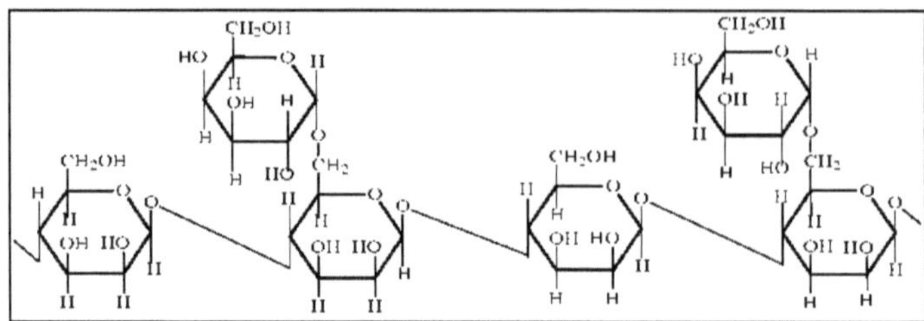

Figure 3.1- Structure of galactomannan obtained from *Delonix regia* seeds Source: adapted from KAPOOR, 1972.

There are reports in the literature on the use of gum extracted from flamboyant seeds to microencapsulate nutraceutical substances in foods and medicines (MOROCHI *et al.*, 1999). In one study, Adetogun and Alebiowu (2009) used the species' gum as an encapsulating agent for paracetamol for the controlled release of the drug. Tekade and Gattani (2009) also used the plant's gum to encapsulate theophylline as a controlled release system for the drug in order to offer convenience to chronic asthma patients.

In order to promote greater stability of ascorbic acid (vitamin C), De Souza et al. (2015) used *D. regia* gum to encapsulate the vitamin using the *spray drying* technique. The microparticles obtained showed their potential for use in different product forms, i.e. tablets, gels, sprays, etc. In addition, ascorbic acid encapsulated with galactomannan extracted from *D. regia* can be applied in the cosmetics and pharmaceutical industries and in the food industry as a protective agent against food oxidation and as an antioxidant in the human diet (De Souza et al., 2015).

De Farias and colleagues (2018) also used the species' biopolymer to encapsulate riboflavin (vitamin B2). Vitamin B2 is highly photosensitive and is widely used as a nutraceutical substance (AHMAD et al., 2004; MOFFAT et al., 2011). The results obtained showed that the vitamin encapsulated with *D. regia* gum can provide subsidies for the preparation of medicines and/or food matrices at low cost and retaining their functional properties (DE FARIAS et al, 2018).

In addition, Frota et al. (2018) encapsulated allantoin with galactomannan extracted from the D. regia species using the spray drying atomization technique for controlled release. The controlled release of the microcapsules and of the active ingredient allantoin alone showed results in accordance with the Brazilian Pharmacopoeia 2010, showing a gradual release of the allantoin present in the microparticles over the time stipulated for analysis compared to the release of the pure active ingredient, which was very rapid (FROTA et al., 2018). This result allowed us to infer that the production of these microparticles, in addition to promoting the protection of the active ingredient, increasing its stability and useful life, also allows this material to be administered as a drug for typical use in patients who need skin cell regeneration treatment (FROTA et al., 2018).

Although there are few studies, this gum can be considered useful for microencapsulation, as it contains galactomannan-type polysaccharides, similar to those in guar gum (*Cyamopsis tetragonalobus*) and locust bean gum (*Prosopis chilensis*), which have already been studied extensively as encapsulating agents (MATSUHIRO *et al.*, 2006).

REFERENCES

ADETOGUN, G. E. & ALEBIOWU, G. B. E. N. G. A. Properties of *Delonix regia* seed gum as a novel tablet binder. **Acta Poloniae Pharmaceutica**. v.66, p.433-438, 2009.

AHMAD, I., FASIHULLAH, Q., NOOR, A., ANSARI, I. A., & ALI, Q. N. M. Photolysis of riboflavin in aqueous solution: a kinetic study. **International journal of pharmaceutics**. v. 280, p.199-208, 2004.

AL-DEGS, Y. S.; EL-BARGHOUTHI, M. I.; EL-SHEIKH, A. H. & WALKER, G. M. Effect of solution pH, ionic strength, and temperature on adsorption behavior of reactive dyes on activated carbon. **Dyes and pigments**. v. 77, p. 16-23, 2008.

AQIL, F.; KHAN, S.M.A.; OWAIS, M. & AHMAD, I. Effect of certain bioactive plant extracts on clinical isolates of beta-lactamase producing methicillin resistant *Staphylococcus aureus*. **Journal of Basic Microbiology**. v. 45, p. 106-114, 2005.

AZERO, E. G. & ANDRADE, C. T. Extraction and characterization of the galactomannan from the seeds of *Caesalpinia pulcherrima*. **Polimeros**, v. 9, p.54-59, 1999.

AZERO, E.G. & ANDRADE, C.T. Testing procedures for galactomannan purification. **Polymer Testing**. v.21, p. 551-556, 2002.

BAVEJA, S.K.; RAO, K.V.R.; ARORA, J.; MATHUR, N.K. & VINAYAH, V.K. Chemical investigations of some galactomannan gums as matrix tablets for sustained drug delivery. **Indian Journal of Chemistry**. v.30, p. 133-137, 1991.

BENTO, J. F.; MAZZARO, I.; DE ALMEIDA SILVA, L. M.; DE AZEVEDO MOREIRA, R.; FERREIRA, M. L.; REICHER, F. & DE OLIVEIRA PETKOWICZ, C. L. Diverse patterns of cell wall mannan/galactomannan occurrence in seeds of the Leguminosae. **Carbohydrate Polymers**. v. 92, p. 192-199, 2013.

CUNHA, P. L.R.; VIEIRA, I. G.P.; ARRIAGA, A. M.C.; PAULA, R.C.M. & FEITOSA, J. P.A. Isolation and characterization of galactomannan from *Dimorphandra gardnerianatul.* seeds as a potential guar gum substitute. **Food Hydrocolloids**. v.23, p.880-885, 2009.

DA SILVA, L. B. **Use of adsorbents from rice husks to remove copper from aqueous effluents.** 2013. 110 f. Dissertation (Master's Degree) - Doctoral Course in Chemistry, FEDERAL UNIVERSITY OF BAHIA, Salvador, 2014.

DAUD, W. M. A. W. & Ali, W. S. W. Comparison on pore development of activated carbon produced from palm shell and coconut shell. **Bioresource Technology**, v.93, p.63-69, 2004.

DEA, I.C.M. & MORRISON, A. Chemistry and Interactions of Seed Galactomannans. **Advances in Carbohydrate Chemistry and Biochemistry**, v.31, p.241-312, 1975.

DE FARIAS, S. S.; SIQUEIRA, S. M. C.; CUNHA, A. P.; DE SOUZA, C. A. G.; DOS SANTOS FONTENELLE, R. O.; DE ARAÙJO, T. G.; DE AMORIM, A. F. V.; DE MENEZES, J. E. S. A.; DE MORAIS, S. M. & RICARDO, N. M. P. S.

Microencapsulation of riboflavin with galactomannan biopolymer and F127: Physicochemical characterization, antifungal activity and controlled release. **Industrial crops and products**. v. 118, p. 271-281,2018.

DE SOUZA, C. F.; LUCYSZYN, N.; FERRAZ, F. A. & SIERAKOWSKI, M. R. *Caesalpinia ferrea var. ferrea* seeds as a new source of partially substituted galactomannan. **Carbohydrate Polymers**. v.82, p.641-647, 2010.

DE SOUZA, C. A.; SIQUEIRA, S.; DE AMORIM, A. F.; MORAIS, S. M. D.; GONÇALVES, T.; GOMES, R. N. & RICARDO, N. M. Encapsulation of l-ascorbic acid within the natural biopolymer-galactomannan-using the spray-drying method: preparation, characterization, and evaluation of antioxidant activity. **Quimica Nova**, v. 38, p. 877-883, 2015.

DE SOUZA, C. A. G. **Encapsulation of ascorbic acid: An alternative to increase its stability**. 1. ed. Mauritius: Novas Ediçôes Acadêmicas, 2018. v. 1. 89p.

DIAO, Y.; WALAWENDER, V. P. & FAN, L. T. Activated carbons prepared from phosphoric acid activation of grain sorghum. **Bioresource Technology**, v. 81, p.4552, 2002.

FROTA, H. B. M.; MENEZES, J. E. S. A.; SIQUEIRA, S.M.C.; RICARDO, N. M. P. S.; ARAÙJO, T.G.; DE SOUZA, C.A.G.; BANDEIRA, P.N. & DOS SANTOS, H.S. Preparation, physico-chemical characterization and controlled release of galactomannan microparticles containing allantoin. **Quimica nova**. v. 41, p. 544-549, 2018.

GANTER, J.L.M.S.; HEYRAND, A.; PETKOWICKZ, C.L.O.; RINAUDO, M. & REICHER, F. Galactomannan from Brazilian seeds: characterization of the oligosaccharides produced by mild acid hydrolysis. **International Journal of Biological Macromolecules**, v.17, p. 13-19, 1995.

GIDLEY, M.J. & REID, J.S.G. Galactomannans and other cell wall storage polysaccharides in seeds. **Food polysaccharides and their applications, 2nd edn. Taylor & Francis, Boca Raton, FL**. p. 181-215, 2006.

GIRGIS, A. S.; YUNIS, S. S. & SOLIMAN, A.M. Characteristics of activated carbon from peanut hulls in relation to conditions of preparation. **Materials Letters**, v.57, p.164-172, 2002.

GONZALEZ, M. P. E. & MONTOYA, V. H. Guava seed as an adsorbent and as a precursor of carbon for the adsorption of acid dyes. **Bioresource Technology**, v.100, p.2111-2117, 2009.

GRECCO, S. T. F.; DE SOUZA, C. A.; BRAGA, G.; R. M.; MELO, D. M. A.; VASQUEZ, D. P. R. & RANGEL, M. C. Evaluation of USY-supported gallium in the fast pyrolysis of Flamboyant pods. In: INTERNATIONAL ZEOLITE CONFERENCE - ZEOLITES FOR A SUSTAINABLE WORLD, 18., 2016, Rio de Janeiro. **Proceedings....** Rio de Janeiro: International Zeolite Conference - Zeolites for A Sustainable World, 2016. p. 1 - 2.

HAMEED, B. H.; DIN, AT MOHD & AHMAD, A. L. Adsorption of methylene blue onto bamboo-based activated carbon: kinetics and equilibrium studies. **Journal of hazardous materials**. v. 141, p. 819-825, 2007.

HAYASHI, J., HORIKAWA, T., TAKEDA, I., MUROYAMA, K. & ANI, F. N. Preparing activated carbon from various nutshells by chemical activation with K2CO3. **Carbon**, v.40, p.2381-2386, 2002.

JUAN, Y.& QIANG, Q. Preparation of Activated Carbon by Chemical Activation under Vacuum, **Environmental Science & Technology**, v.43, p.3385-3390, 2009.

KALDERIS, D.; KOUTOULAKIS, D.; PARASKEVA, P.; DIAMADOPOULOS, E.; OTAL, E.; VALLE, E. O. D. & PEREIRA, C. F. Adsorption of polluting substances on activated carbons prepared from rice husk and sugarcane bagasse. **Chemical Engineering Journal**, v.144, p.42-50, 2008.

KAPOOR, V. P. A galactomannan from the seeds of *Delonixregia*. **Phytochemistry**. v. 11, p. 1129-1132, 1972.

KRISHNAIAH, Y.S.R.; KARTHIKEYAN, R.S.; GOURI SANKAR, V. & SATYANARAYANA, V. Three-layer guar gum matrix tablet formulations for oral controlled delivery of highly soluble trimetazidine hydrochloride. **Journal of Controlled Release**. v.81, p. 45-56, 2002.

KUMAGAI, S.; SHIMIZU, Y.; TOID, Y. & ENDA, Y. Removal of dibenzothiophenes in kerosene by adsorption on rice husk activated carbon. **Fuel**, v.88, p.1975- 1982, 2009.

LAPASIN, R. & PRICL, S. **Rheology of industrial polysaccharides- theory and applications**. Gaithersburg: Aspen Publishers, p.620, 1999.

LIMA, S. B.; BORGES, S. M. S.; RANGEL, M. D. C. & MARCHETTI, S. G. Effect of iron content on the catalytic properties of activated carbon-supported magnetite derived from biomass. **Journal of the Brazilian Chemical Society**, v. 24, p.344-354, 2013.

LIMA, S. B. **PREPARATION OF ACTIVATED CHARCOAL FROM GREEN COCONUT MESOCARP USING DIFFERENT ACTIVATING AGENTS.** 2013. 140 f. Thesis (Doctorate) - Doctorate in Chemistry, Federal University of Bahia, Salvador, 2013.

LOPES, G.C.; MACHADO, F.A.V.; TOLEDO, C.E.M.; SAKURAGUI, C.M. & MELLO, J.C.P. Chemotaxonomic significance of 5-deoxyproanthocyanidins in Stryphnodendron species. **Biochemical Systematics and Ecology**, v. 36, p. 925-931, 2009.

MATSUHIRO, B.; LILLO, L.E.; SAENZ, C.; URZÙA, C.C. & ZARATE, O. Chemical characterization of the mucilage from fruits of *Opuntiaficus-indica*. **Carbohydrate Polymers**. v. 63, p. 263-267, 2006.

MCCLEARY, B.V.; DEA, I.C.M.; WINDUST, J. & COOKE, D. Interaction properties of D-galactose-depleted guar galactomannan samples. **Carbohydrate Polymers.** v. 4, p. 253-270, 1984.

MERCÊ, A.L.R.; LOMBARDI, S.C.; MANGRICH, A.S.; REICHER, F.; SZPOGANICZ, B. & SIERAKOWSKI, M.R. Equilibrium studies of galactomannan of *Cassia fastuosa* and *Leucaena leucocephala* and Cu^{2+} using potentiometry and EPR spectroscopy. **Carbohydrate Polymers**. v.35, p. 13-20, 1998.

MOFFAT, A. C., OSSELTON, M. D., WIDDOP, B., & WATTS, J. **Clarke's analysis of drugs and poisons** (Vol. 3). London: Pharmaceutical press, 2011.

MOROCHI, D.; SAN MARTIN, E. Y & ORDONEZ, D. Estudio sobre la extracción y caracterización de la goma de flamboyàn (Delonix *regia*). **Ciencia y Tecnologia de los Alimentos.** v. 1, p. 59-61, 1999.

NABAIS, J. M. V., LAGINHAS, C. E. C., CARROTT, P. J. M. & CARROTT, M. M. L. R. Production of activated carbons from almond shell. **Fuel Processing Technology**, v. 92, p.234-240, 2011.

NUNES, P.C.; VIVAN, J.L.Forests, agroforestry systems and their environmental and economic services in Juruena-MT. **Juruena Rural Development Association.** Cuiabà,

2011.

OLIVEIRA, S. B.; BARBOSA, D. P.; MONTEIRO, A. P. M.; RABELO, D. & RANGEL, M. C., Evaluation of copper supported on polymeric spherical activated carbon in theethylbenzene dehydrogenation, Catalysis Today, v. 133-135, p. 92-98, 2008.

OWOYOKUN, T. O. Biosorption of methylene blue dye aqueous solutions on *Delonix regia* (flamboyant tree) pod biosorbent. **The Pacific Journal of Science and Technology**. v.10, p.872, 2009.

PANEGASSI, V. R.; SERRA, G. E. & BUCKERIDGE, M. S. Technological potential of galactomannan from faveiro *(Dimorphandra mollis)* seeds for use in the food industry. **Ciência e Tecnologia de Alimentos [online]**. v.20, p. 406-415,2000.

PEREIRA, P.; OLIVEIRA, L. C. A.; VALLONE, A.; SAPAG, K &PEREIRA, M. Preparation of activated carbon at low carbonization temperatures from coffee waste: use of $FeCl_3$ as an activating agent. **Quimica Nova**, v.31, p.1296-1300, 2008.

POLLARD, M.A. & FISCHER, P.A. Partial aqueous solubility of low-galactose-content galactomannans - What is the quantitative basis? **Current Opinion in Colloid & Interface Science**, v. 11, p. 184-190, 2006.

PONNUSAMI, V.; GUNASEKAR, V. & SRIVASTAVA, S. N. Kinetics of methylene blue removal from aqueous solution using gulmohar *(Delonix regia)* plant leaf powder: multivariate regression analysis. **Journal of hazardous materials**. v.169, p.119-127, 2009.

PRAJAPATI, V.D.; JANI, G.K.; MORADIYA, N.G.; RANDERIA, N.P.; NAGAR, B.J.; NAIKWADI, N.N. & VARIYA, B.C. Galactomannan: A versatile biodegradable seed polysaccharide. **International Journal of Biological Macromolecules**, v. 60, p.83-92,

2013.

RAMAKRISHNAN, M. & NAGARAJAN, S. UTILISATION of Flame Tree Waste Biomass for the Removal of Hg (II) from Water. **Acta Chimica Slovenia**. v.56, p.282-287, 2009.

RECHIA, C.G.V.; SIERAKOWSKI, M.R.; GANTER, J.L.M.S. & REICHER, F. Polysaccharides from the seeds of *Senna multijuga*. **International Journal of Biological Macromolecules**. v.17, p. 409-412, 1995.

RODRiGUEZ-REINOSO, F. & SEPÙLVEDA-ESCRIBANO, A. Porous Carbons in Adsorption and Catalysis. **Handbook of Surfaces and Interfaces of Materials**. Chapter 9, v.5, p.309-355, 2001.

STEPHEN, A. M. & CHURMS, S. C. **Gums and mucilages**. In: Stephen, A. M. (Ed) Food Polysaccharides and their applications. New York: Marcel Dekker. p. 377440, 1995.

SUGUMARAN, P.; SUSAN, V. P.; Ravichandran, P. & SESHADRI, S. Production and characterization of activated carbon from banana empty fruit bunch and *Delonix regia* fruit pod. **Journal of Sustainable Energy & Environment**. v.3, p.125-132, 2012.

TEKADE, A. R. & GATTANI, S. G. Development and evaluation of pulsatile drug delivery system using novel polymer. **Pharmaceutical development and technology**. v. 14, p.380-387, 2009.

VALLILO, M.I.; CARUSO, M.F.S.; TAKEMOTO, E.; PIMENTEL, P.A. Chemical and physico-chemical characterization of the oil from the seeds of *Platymiscium floribundum* Vog. (sacambu), harvested at the stage of development and at the time of physiological maturation. **Revista do Instituto Florestal**, v.19; p.73-80, 2007.

VARGAS, A. M.; CAZETTA, A. L.; GARCIA, C. A.; MORAES, J. C.; NOGAMI, E. M.; LENZI, E. & ALMEIDA, V. C. Preparation and characterization of activated carbon from a new raw lignocellulosic material: Flamboyant (*Delonix regia*) pods. **Journal of environmental management.** v.92, p.178-184, 2011.

VARGAS, A. M.; CAZETTA, A. L.; KUNITA, M. H.; SILVA, T. L. & ALMEIDA, V. C. Adsorption of methylene blue on activated carbon produced from flamboyant pods (*Delonix regia*): Study of adsorption isotherms and kinetic models. **Chemical Engineering Journal.** v.168, p.722-730, 2011.

VARGAS, A. M.; CAZETTA, A. L.; MARTINS, A. C.; MORAES, J. C.; GARCIA, E. E.; GAUZE, G. F. & ALMEIDA, V. C. Kinetic and equilibrium studies: Adsorption of food dyes Acid Yellow 6, Acid Yellow 23, and Acid Red 18 on activated carbon from flamboyant pods. **Chemical Engineering Journal.** v.181, p.243-250, 2012.

VARGAS, A. M.; GARCIA, C. A.; REIS, E. M.; LENZI, E.; COSTA, W. F. & ALMEIDA, V. C. NaOH-activated carbon from flamboyant (*Delonix regia*) pods: optimization of preparation conditions using central composite rotatable design. **Chemical Engineering Journal.** v.162, p. 43-50, 2010.

VARSHOSAZ, J.; TAVAKOLI, N. & ERAM, S.A. Use of natural gums and cellulose derivatives in production of sustained release metoprolol tablets. **Drug Delivery.** v.13, p. 113-119, 2006.

VENDRUSCOLO, C. W.; FERRERO, C.; PINEDA, E. A.; SILVEIRA, J. L.; FREITAS, R. A.; JIMÉNEZ-CASTELLANOS, M. R. & BRESOLIN, T. Physicochemical and mechanical characterization of galactomannan from *Mimosa scabrella*: Effect of drying method. **Carbohydrate Polymers.** v.76, p.86-93, 2009.

VIEIRA, í. G. P.; MENDES, F. N. P.; GALLÂO, M. I. & DE BRITO, E. S. NMR study of galactomannans from the seeds of mesquite tree (*Prosopis juliflora* (Sw) DC). **Food chemistry.** v. 101, p.70-73, 2007.

YANG, T. & LUA, A. C. Characteristics of activated carbons prepared from pistachio-nut shells by potassium hydroxide activation. **Microporous and Mesoporous Materials**, v.6, p.113-124, 2003.

YAVUZ, R.; AKYILDIZ, H.; KARATEPE, N. & ÇETINKAYA, E . Influence of preparation conditions on porous structures of olive stone activated by H_3PO_4. **Fuel Processing Technology**, v.91, p.80-87, 2010.

ZABIHI, M.; ASL, A. H. & AHMADPOUR, A. Studies on adsorption of mercury from aqueous solution on activated carbons prepared from walnut shell. **Journal of Hazardous Materials**, v.174, p.251-256, 2010.

ZANI, L. B.; DUARTE, I. D.; MOROZESK, M.; BONOMO, M. M.; ROCHA, L. D. & CORTE, V. B. The use and potential of forest seeds. **Natureza on line**, v.11, p.118-124, 2013.

Printed by Books on Demand GmbH, Norderstedt / Germany